Water Wells

What a Dowser
Needs to Know

Susan Collins

Water Wells
What a Dowser Needs to Know
Susan Collins

First Edition: June, 2009, ISBN: 978-0-9780899-6-2
This edition: June, 2022 Published by: Susan Joan Collins
335 Patricia Drive, King City, Ontario, L7B 1H4, Canada

Check the other Kindle books by Susan Collins in the Complete Guide to Dowsing series on Amazon and on www.dowser.ca. (See back pages for titles)

Rent dowsing workshops from
https://vimeo.com/susancollinsdowser/vod_pages

To order books and tools, or arrange a workshop or
personal session contact Susan at:
susan@dowser.ca
www.dowser.ca

Table of Contents

2. **All About Water** … 9
      The water cycle … 9
      Water consumption … 10

3. **All About Wells** … 13
      Types of wells … 13
      Drillers … 16
      Locating a well … 17
      Maintaining a well … 19
      Water testing and treatment … 20

4. **All About Dowsing** … 25
      Tools … 26
          Pendulum … 62
          L-rods … 28
          Y-rod … 30
          Body… 31
          PLUS / MINUS Chart … 31
      The Dowsing Protocol … 34
      Belief Systems … 39

5. **Dowse a Water Well** … 33
      Triage Protocol … 43
      Map Dowsing … 47
      Stream Diversion … 49

6. **Conclusion** … 51

      Glossary … 52
      About the author … 57

*Thank you to King City Well Drilling
for all the on-the-job training!*

1.   Introduction

Back before hydrogeology was a science and there were even such things as Well Records, people who needed a new well had a tried-and-true way of figuring out where to dig for water. Most rural communities had a "Witcher", someone who would come out to a property, cut a forked branch from a supple tree, and walk the land until the branch dipped down, showing the best place to find a water supply. Witchers tended to refer to their ability as a special gift, implying that it couldn't be learned, that one had to be born with it. (By the way, the term *"witch"* comes from the Old English word *"wych"* meaning "pliant" which refers to the bendable dowsing stick used.)

Over time, as modern technology replaced the old ways, water dowsers (as Witchers came to be known) became harder to find, and these days, it is rare to meet one. As a past president of the Canadian Society of Dowsers (CSD), I know that to be true from the look of incredulity on peoples' faces when I tell them what I do for a living.

People are even more amazed when I tell them that modern dowsers don't just find the energy of water and minerals, but that we also work with the energy of plants, animals and the human body. Not only that, that anyone with desire and patience can be taught to dowse by learning to focus their mind and listening to their body.

By the way, the Glossary at the back of the book may help you with terms you aren't familiar with.

How I Became a Dowser

I became a dowser by accident. I was diagnosed with Rheumatoid Arthritis way back in 1985 and had trouble controlling my symptoms with either western or alternative medicines. I read about dowsing as a way of improving health and took a course through the CSD. (My health did improve dramatically and today I am virtually symptom and drug free.)

Back to my first well in 2005: one day I got a call from someone who needed a well, and I was the only person available to do the location. I was nervous, but went on site, did what I'd been taught, and low and behold, the water came in at the place, depth and volume that I'd said it would. After that experience, I began dowsing wells fairly regularly, and have a good track record. Over the years I've worked with a few drillers, and found them all to be an open-minded group, many of whom are closet dowsers.

Why Use a Dowser for Well Location?

Using a dowser at a drill site can increase the likelihood of finding a good well with the first hole. Sometimes dowsers are called in after a driller hasn't met with success. A driller might call in a dowser to give a second opinion if he has a client that unreasonably expects to have a hundred Gallon-Per-Minute (gpm) gusher at his back door in an area that is known to usually provide less than ten gpm. Sometimes clients call in a dowser for an assessment of the property even before they call the driller.

I've worked with people at an undeveloped property while they were considering an offer to purchase. I have worked at other places with the architect before the house is sited to be sure that there is water where the homeowner will need it. The worst-case scenario I've worked at is where the house was already built, the wiring was in the ground, and they had left a six-by-six meter patch of dirt to find the water they needed.

How Does Dowsing Work?

Moving water, even when it's underground, causes electricity to flow by raking off electrons. Our bodies have natural electromagnetic receptors, so with practice we can learn to "feel" water below the ground. It is perhaps similar to our ability to sort out different audio frequencies to identify different types of music. Dowsers use tools to amplify these subtle signals in their bodies.

What are Dowsing Tools?

Anything that moves can be a dowsing tool: Y-rods (forked sticks); pendulums (any weighted object on a string); L-rods (bent wires); even your fingers. But the tools themselves have no magic power, just as a hammer has no power to drive a nail until we pick it up. The tools simply amplify our body's sensory perceptions.

One thing I do, which perhaps makes me look a little odd in the field, is talk to my tools. I'll say, for example, *"L-rods, point to the best place to drill for water on this site."* It's not that I think the tools can hear me, but addressing them helps me focus my thoughts, to make sure my mind-body connection is completely centered. The key to all successful dowsing is to make sure your thoughts are focused, and that your questions are clear. It is also important to stay emotionally detached from the outcome, even though you know it is important to find water, and expensive to make a mistake.

The Water Cycle

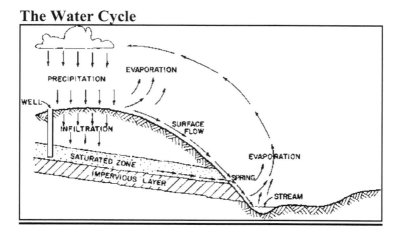

Wells are developed in saturated zones in order to access water moving through the Water Cycle.

2. All about Water

The Water Cycle

The water cycle is the movement of water in the environment. Water falls to the earth as precipitation, either rain or snow, then soaks into the earth or runs into streams, lakes and rivers. The water that soaks into the earth is known as groundwater. Moisture evaporates from surface water bodies, and eventually condenses into clouds that repeat the cycle, falling to earth again as rain or snow.

Groundwater

Much of the world's fresh water is found as groundwater in aquifers, the underground areas of soil or rock that hold water either in cracks (in bedrock formations), or in the tiny spaces between soil particles (in a gravel formation.)

Wells are drilled into aquifers to access groundwater. The top of the water in these aquifers is called the water table. Groundwater is generally safer than surface water for drinking because of the filtration and natural purification processes that take place in the ground.

Generally, aquifers are not as easily contaminated as surface water, but once they are contaminated, it is much more difficult to clean them. They can be contaminated when household, commercial, or industrial wastes such as sewage, fertilizers, toxic chemicals, and road salts, seep into the ground. Acid rain can also contaminate aquifers through the processes of the water cycle.

Water Consumption

You'll need to determine how much water your client needs from the well. In 1998, Canadians, on average, used 343 litres (about 91 U.S. gallons) of water per person per day. Well yields are referred to as gallons per minute (gpm) or litres per second (l/sec). Residential uses typically account for about half of overall water usage, while about a quarter is used by commercial and industrial uses. More than ten percent of municipal water is typically lost due to leakage.

There is often a basic minimum water yield a well must produce in order to qualify for a mortgage. Usually, the lowest viable yield is 5 gpm. If the well is not expected to produce sufficient volume for direct delivery into the home, a cistern can be installed to gradually build up reserves for later use.

If the end user of the well is commercial, the business owner should have a good idea of how much water they need. A client of mine who has a riding stable with many horses as well as cows in pasture surprised me by telling me he runs his home and operation on a well producing only six gpm.

Typical Residential Use

Fixture or water use	With regular fixtures you use:	With water-saving fixture you use:
Shower	7 gallons per minute	2.5 gallons per minute
Bath	30 gallons for full tub	14-18 gallons for 1/2 tub
Running faucet	3 gallons per minute	2 to 2.5 gpm / minute
Washing machine	40-55 gallons per load	18-25 gallons a load
Dishwasher	15 gallons per load	
Toilet	6 gallons per flush	1.6 gallons per flush

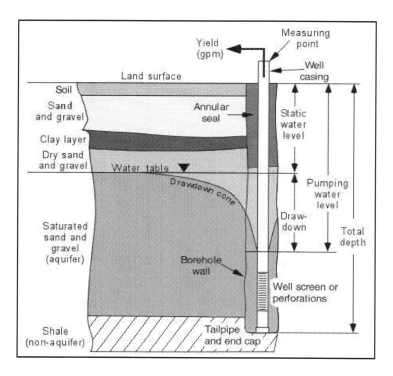

Wells safely penetrate aquifers to provide sustainable water supplies. See Glossary for definitions of terms.

3. All about Wells

Types of Wells

A well is a hole is drilled or dug into an underground water source (aquifer), that provides water at the surface. Once the well has been drilled, a submersible pump is installed into the casing that can pump the water up on demand. Normally the water is pumped into a water pressure tank and then is available for use in the house. Sometimes purification systems are installed on the system if water testing indicates that is necessary. Cisterns can hold reserves.

The yield of a well depends on the aquifer, the placement of the well and the depth and nature of the well itself. The geology and climate of the area determine the availability of water in the aquifer. An area of low rainfall will likely have deep aquifers. Where the geology is bedrock, the aquifers will likely be contained in fissures in the rock. In a moraine geology, the water will percolate through the sands and gravels.

No one can guarantee what the yield will be in any given location, but local Well Records can serve as a rule of thumb. You can begin to estimate the potential yield of a new well by talking to neighbours, or the people at the local building supply store to get an idea of well yields in the area.

The type of water well installed at a site depends on the local geology, the water requirements of the client, and local government regulations. Most modern water wells are drilled, but some are also dug, driven, jetted or bored (augered). Geothermal wells, used for heating and cooling purposes, are becoming more common.

Wells are lined with casings (usually either steel or PVC plastic) to keep the hole open and permit installation of the necessary pumping or injection equipment. The casing should extend above the ground. Overburden (sand and gravel) wells have casing extending the entire depth of the well to keep the borehole open and to seal off any surface water. Bedrock wells do not usually require casing in the rock itself.

The casing is screened or open at intervals in the casing (usually at the bottom) to allow water to enter the well casing and keep sand or other debris out. There should always be a seal around the well casing to prevent surface contamination of the aquifer.

Drilled Wells

Most modern wells are drilled, which requires a rig, often mounted on a big truck. Rigs use rotary drill bits that "chew" at the rock, percussion bits that smash the rock, or, if the ground is soft, large auger bits to pull up the ground. Drilled wells are typically about fifteen cm (six inches) in diameter.

Dug Wells

Most hand-dug wells are relics from a time before drilling equipment was readily available. They are rarely installed now (and prohibited in many areas) because of the high potential for either contamination or for running out of water.

Dug wells have a large diameter, typically one meter (three feet), and are lined with concrete tiles, galvanized steel culverts, or fibreglass. Some may have the wellhead located below ground level, which is no longer considered a safe practice as they often collect debris.

Driven and Sand Point Wells

These are small diameter wells, typically five cm (two inches) or less, where screens and pipes are driven into sandy soils in an area with a high-water table. They are very susceptible to water contamination because of their shallow depths.

Geothermal Wells

Geothermal wells use the temperature of the earth to heat or cool buildings using a heat exchange principle. A circulating pump delivers the water (or other eco-friendly solution) from the system to the boiler and back to the system in a continuous closed loop resulting in a heat transfer.

Deep geothermal wells or "loops" can bring heat up from the earth's crust, while horizontal loops can make use of the ambient underground temperature to moderate the building temperature. Small, residential geothermal wells are relatively rare in most areas because it is often uneconomical to drill a well of more than a few hundred feet to serve the domestic needs of a single home.

There are several different well configurations depending on the land available for installation:

Horizontal loops are often used when adequate land surface is available. Pipes are placed in trenches, in lengths that range from 50 to 400 meters (150 to 1,200 feet).

Vertical loops are used when available land surface is limited. Drilling equipment is used to bore small-diameter holes from 50 to 200 meters (150 – 600 feet) deep.

Pond (lake) loops are very economical to install when a body of water is available, because excavation costs are virtually eliminated. Coils of pipe are simply placed on the bottom of the pond or lake where the water temperature remains a constant throughout the year.

Open loop systems utilize ground water as a direct energy source. Two boreholes are drilled, one acts as a supply borehole, the other as an injector well. Sometimes called a "pump and dump" system, there is growing concern that this type of system can contaminate the groundwater.

Drillers

Well drillers are licensed contractors, who should have special training in constructing wells. In Ontario,

Canada where I live, they are regulated by the Ministry of the Environment as set out in Regulation 903. Industry associations include the Ontario Groundwater Association (OGWA), the Canadian Ground Water Association, and The National Drilling Association (USA). Check the levels of government in your area to see who regulates well drilling.

Sometimes the best way to find a reliable, local driller is just to ask your neighbours or the manager at the local hardware store. Here are some things to consider when choosing a driller:

- Do they have proof of licensing and insurance?
- Do they provide a written estimate?
- What is included in the price?
- What is the proposed construction method?
- Are they aware of regulatory requirements?
- Do they provide a Well Record?
- Do they provide any guarantees?
- Are they a full-service company or do they subcontract some of the work?
- What is the proposed maintenance schedule?
- Do they clean-up after the job?
- Do they have local references?
- What is the time frame to completion?

Locating a Well

The location of a well depends on the geology of the area and the equipment available to construct it. Drillers rely on their local knowledge of an area to site the well. They may refer to hydrogeologists' maps that mark underground water, or the Well Records from

17

previously developed wells. Some drillers may also dowse a location themselves or call in a dowser to assist.

All wells should be easy to install, maintain and placed away from sources of contamination. If a new well is installed, most jurisdictions require the old one to be properly sealed by a licensed contractor.

The Physical Structure of the well

- The well casing should be intact and extend above the ground.
- The seal around the casing should be tight.
- The ground surface around the top of the well should be sloped up towards the top of the casing.
- The well casing cap should be solid and fit tightly.
- Water should not seep along the inside of the casing above the water level in the well.
- There should be no biological material such as animals, insects or roots in the well.
- The electrical wiring to the well should be enclosed in a conduit between the well and ground, and between the basement wall and the pressure switch.

Distance from contaminants
In Ontario, Canada, provincial legislation dictates that a drilled well must be at least 50 feet from any potential source of contamination, including the septic tank and tile bed. Other jurisdictions will have similar requirements.

Plumbing accessibility

Try to have the well and the utility room on the same side of the house to minimize trenching and underground pipes.

Environmental issues

Make sure the new well does not interfere with other wells in the area or is near contaminant sources.

Maintaining a Well

Water Treatment

Installing a water-conditioning unit can reduce water hardness and iron content but does not solve contamination issues. Other water quality problems may sometimes be solved through disinfection of the well water distribution system.

Maintenance Accessibility

It is important to maintain a well to preserve the quality of water in your own system and in the aquifer. Wells and equipment must be sited so they can be easily accessed at all times for cleaning, treatment, repair, testing, and visual examination. Access from a driveway and at least ten feet from a structure or tree is a good rule of thumb. Make sure there will be no driveway run-off towards the well. A properly constructed well forms an effective barrier against surface run-off that may enter and contaminate the well.

Operating Issues

Well pump turns on and off continuously when used. A waterlogged pressure tank may cause this condition. Draining and re-pressurizing the pressure tank may correct the problem.

Poor water pressure. This may be caused by an improper pump setting on the pressure switch. Most pressure switches are set to turn the well pump on at 30 p.s.i. (pounds per square inch) and off at 50 p.s.i.

The well pump turns on when water is in use. There may be a leak somewhere in the water system. Check the inside plumbing for leaks and outside for wet spots in the yard.

Water Testing and Treatment

The quality of well water depends on the type of well and the area. Wells should be tested for total coliform bacteria and E. coli prior to being used, and at least semi-annually thereafter. In Ontario, Canada, County Health Units provide bacteriological testing free of charge.

You can also test for chemicals such as pesticides and nitrates, as well as iron, sulphur, hardness, salt, nuisance bacteria such as iron-related bacteria. These tests may need to be done by a private lab on a fee for service basis.

It is important to regularly test the water and inspect and maintain the well in order to avoid problems with the water system. Symptoms of poor water quality include:
- the presence of coliform bacteria
- changes in the chemical quality of the well water as detected through laboratory analysis.

- changes in the quality of the water such as turbidity (cloudiness), colour, taste or odour.
- rapid or large changes in the water level in the well.

Disinfecting Your Water Well
Shock chlorination is used to disinfect the well, pump, and piping to get rid of coliform and E. coli bacteria and to reduce or control iron-related and sulphur reducing bacteria.

Wells should be chlorinated after drilling, after pump installations, and if the water test indicates the presence of undesirable elements.

Although chlorinating is safe for the well components, chlorine is corrosive and can be harmful to you if it is mishandled. Make sure that you have good ventilation when using bleach. Use this procedure at your own risk or ask a licensed well technician to do it for you.

The water supply is not considered safe until a satisfactory laboratory report has been received after the disinfection procedure. Until you get a safe test back, you should boil all drinking water for five minutes or use bottled water. Water may also be made safe for drinking by putting five drops of unscented bleach into each gallon of water. Let the water stand for thirty minutes before drinking. This method should be used only with water that is clean in appearance and free of odour.

A Procedure for Disinfecting a Well
Even though a disinfection method is given below, it is not guaranteed, so check with your local health unit or

driller for specific recommendations for your area and your type of well. You are responsible for your actions.

1. Put water treatment units on "bypass" as chlorine may damage them. Do not get water or chlorine into any electrical connections. Disinfect water treatment equipment by using the manufacturers' suggested procedure.

2. Determine the amount of bleach you will need for the well by looking up the depth of the water in your well in your Well Record and checking the table, below. The water depth is calculated as the total depth minus the static water level. For the average home well, one to two gallons of bleach will be adequate. Add another litre of bleach to account for the volume of the hot water tank and pressure tank.

3. Use ordinary, (non-scented) household bleach containing chlorine.

Depth of water in existing well (1 metre = 40 inches)	Volume of 5.25% bleach added	
	Casing Diameter 15 cm (6 inches) drilled	**Casing Diameter** 90 cm (35 inches) dug
3 metres	60 ml 2 oz	.6 Litres .5 qt
5 metres	100 ml 3.5 oz	3 Litres 2.6 qt
10 metres	200 ml 7 oz	6.5 Litres 5.7 qt
New wells require a chlorine concentration of 250 parts per million (ppm). Existing wells require 50 ppm chlorine. Health Canada		

4. Mix the bleach in a bucket with three or four gallons of water. You can use the well water.

5. If possible, **circulate water through the system** by running water into the well from a hose connected to an outdoor tap.

6. While the water is running, **add the chlorine solution** to the well by slowly trickling it in, or by adding small amounts at intervals. If you have a well seal, you may add the chlorine through the air vent using a funnel.

7. Run the hose in the well for 15 to 20 minutes after you first notice a chlorine smell coming from the water in the garden hose, and then shut off the garden hose. This will help to get a uniform concentration of chlorine in the well and pressure system.

8. Remove the hose from the well and close the well cap.

9. Turn on each water faucet serviced by the well (including the washing machine and dishwasher) one at a time until you smell chlorine, then shut it off. Flush toilets.

10. Allow the bleach solution to remain in the water lines for at least two hours. (Don't use your taps!) Then run each tap for ten seconds, close them, and let them stand for at least eight hours, or overnight. The water should not be used except for flushing toilets as necessary. **Do not leave the bleach solution in the system for longer than twelve hours**.

11. After above, **connect a hose to an outside water faucet** and run the water into a ditch until the bleach odour disappears. **Do not run water inside as the**

large amount of chlorine may affect the septic system. Do not add this water to a wetland or ditch that connects to a natural waterway. The chlorine may also cause scale and rust to break free from the well, which could plug up your inside plumbing system if you were to turn on the inside taps. When the outside water is odourless, run each inside tap being served by the well until the water is odourless.

12. After two days, and if the odour of bleach is not detected, **re-test your water**. (You should also have your water retested about two weeks after chlorinating the system to makes sure the problem is eliminated.)

13. Once the water is running clear, **use a hose to flush the inside of the well again** to get rid of any chlorine and iron residue that may be on the inside of the well casing. When the water from the hose gets dirty, discharge it away from the well. Repeat this step until the water no longer gets dirty.

14. Flush the rest of the fixtures in the house. Don't do white laundry for two or three loads.

15. Have the water tested to make sure it is safe before using.

4 All About Dowsing

I have found that the easiest and most effective way to access your intuition to find water is to dowse. Dowsing is the detection and transformation of energy with simple tools and the power of heart and thought. It has always been known as a way to find water and minerals, but today modern dowsers also use their skills to detect, and interact with physical, mental, emotional and spiritual energies.

The ability to dowse is a natural, sensory ability we all have. It is not based on a religious practice, but I find my best results come when I align myself with the "Best and Highest Good of All Creation"

To get the results you want it is important to use a Dowsing Protocol (page 29) every time you dowse. No dowser is one hundred percent accurate all the time, but I guarantee that using this protocol every time you dowse will increase your accuracy. Your level of accuracy can drop off if you are tired, hungry, intoxicated or sick. If your physical state is too bad, you may have to try again when you're feeling better. You can also go back to the Dowsing Protocol and repeat the steps to see if you can bring your accuracy up to a reasonable level.

Tools

Learning to use the tools is easy. Learning to relax and focus your mind at the same time takes practice. Anything that moves can be a dowsing tool. The common tools are pendulums, L-rods, Y-rods and body sensations. The tools amplify our body's sensory perceptions..

How to Dowse with Pendulums

The basic movements of the pendulum are swinging back and forth, (either away from you, or across your body), and swinging in circles (clockwise or counterclockwise). There are other subtle movements that you will begin to recognize as you develop a personal vocabulary.

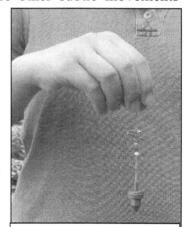

The first step is to determine what YOUR responses mean. Your responses may be different from another person's. This is how to program your responses (and this is the only step where you will consciously be asking the pendulum ·to move in certain directions):

Anything that swings freely can be used a dowsing pendulum.

- Hold the string between your thumb and first finger, about 10 cm (4 inches) from the weight at the end (see photo). While slowly swinging the pendulum back and forth, say out loud *"Pendulum, show me YES"*. The pendulum should move.

- If there is no response, swing the pendulum clockwise for a few seconds (if that is what you'd like your YES response to be), and say: *"Pendulum this is my YES."* Then repeat the question *"Pendulum, show me my YES."* Keep doing this until the pendulum moves on its own. This step creates the brain map that helps your body to amplify subconscious information.

- Repeat these steps for NO, DISCONNECT (used in Step 4 of the Dowsing Protocol, given in a later section) and MAXIMIZE responses (used in Step 5 of the Protocol).

- Practice until you are comfortable with your answers.

What do the different motions mean?

For many people the YES response is a back and forth (towards and away from the body) motion. For others YES is a clockwise, circular spin. NO for many people is a continuous left-right-left swing across the body; for others it is a counter-clockwise spin. Practice until you can reliably recognize YES and NO answers.

For many people the DISCONNECT signal (used in Step 4 of the Dowsing Protocol to remove nonbeneficial energy) is a counterclockwise spin and the MAXIMIZE signal (Step 5, to increase beneficial energy) is a clockwise spin.

Once you recognize your YES and NO responses, let the pendulum give the responses on its own when you ask questions or work with energies. Start the pendulum swinging gently in a neutral direction (an angle slightly off centre), state your intention (or ask your questions), then let the pendulum respond naturally. To find the best place to drill for a water well, ask for well location, depth and volume.

How to Dowse with L-rods

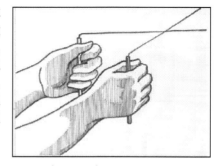

L-rods are the work horses of locating well sites and earth, environmental and living energy fields. They can also be used to determine YES and NO answers.

You can use a coat hanger or bent wire as an L-Rod, or something fancier. It's easiest to work with rods if the handles are made from hollow tubes within which the wires can swing freely no matter how tightly the handles are gripped. Rods made from copper can amplify the dowsing response, making them easy for beginners to use.

Rods can come in any size, from minis that you can keep in your purse or pocket, to larger ones that ultimately may be too heavy and unwieldy, depending on your strength and energy levels.

Addressing the tools by the names *"L-rod!"* or *"Pendulum!")* helps to get things moving. At the end of a session, it's also important to thank the tools, and all levels of consciousness which have contributed to the work.

One student I taught refused to use *"incantations"* with the tools – I told him that stating his intentions out loud simply helped focus his mind / body system. The rods just amplify what you already know – they are not being "controlled" by an outside force.

L-rods show a dowsing response by pointing in certain directions and by swinging back and forth or spinning at different rates. The YES (or "found target") position may be rods open, or rods crossed. Either is fine. Choose one signal and stick with it.

The READY Position: Hold the rod in your fist. With a rod in each hand, and your elbows bent at 90-degree angles, hold the rods pointing away from your body and parallel to the ground with the tips slightly down.

The POINT TO Position: Ask the rods to point to the thing you are looking for, for example the best place to drill to find a water well. The rods will swing in the direction that the target is located. Walk forward in that direction, asking your rods to move to the Found position (below) when you are at the target.

The FOUND Position: Ask the rods to cross when your toes are at the target well site. Confirm you are on the target with a YES or No response.

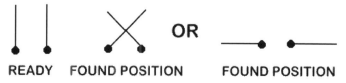

READY   FOUND POSITION   FOUND POSITION

How to Dowse with Y-rods

Y-rods were traditionally made from a pliable forked branch cut from a supple tree. Their name describes their shape. The term *"water witch"* comes from the old English *"wice"* meaning bendable.

READY Position: hold an arm of the Y-rod in each hand, palms up, parallel to the ground. Squeeze your hands together to create tension in the rod, as in the

photo. Turn in a slow circle and ask in which direction is the target. The rod will dip down in that direction.

SEARCHING Position: bring the rod back up so it is parallel to the ground. Walk forward in the direction indicated above, holding the intention that the rod will dip down at the target well site

FOUND TARGET Position: the rod points to the ground. BE CAREFUL! A strong response can smack you "down below". This signal is also used as a NO.

YES Position: the rod points up. A strong response can smack you in the forehead!

How to Dowse with Your Body

There are simple methods for getting a dowsing response just by using your body. Make an "O" with the thumb and fore finger of one hand and try to pull them apart with the thumb and fore finger of the other hand. While you're doing this, ask 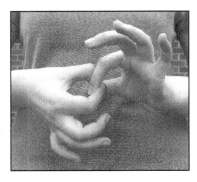 *"Show me YES"*, then *"Show me NO"*. Generally, your YES will be the one that is more difficult to pull apart.

The PLUS / MINUS Chart

If you need to estimate concentrations of dissolved solids, you can use a PLUS / MINUS Chart, see illustration.   To begin, start with your pendulum in neutral, swinging over the centre line that points to zero, then focus on your question *"how much of x is present in a parts per million concentration,* and ask the

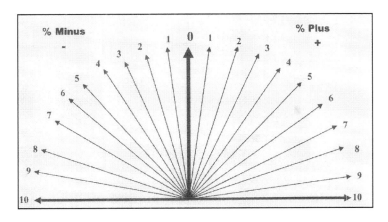

pendulum to swing to the number and line that best reflects the amount. If the pendulum swings to the PLUS 2, for example, you can ask if the reading is .2, 2, 20 or 200 or other variation of the number 2.

The Dowsing Protocol

When you begin as a dowser you might wonder if you're just fooling yourself when the pendulum moves one way or another. You may not trust your answers. The Dowsing Protocol below will guide you through the steps to truly develop your intuition so that you can trust the answers you get with dowsing.

To get accurate results, use a Dowsing Protocol every time you dowse. This practice will help you "zone in" to the energies that are affecting you. It will also provide protection from all sources of nonbeneficial energy and influence so that you can be most accurate.

If you don't use the Protocol or some other sort of energy management and protection system before you dowse, it would be like picking up the phone, dialling random numbers, and taking advice from whoever answered. If you do zone in with a Protocol before you dowse, you'll get your answers from the "Divine Source" (or however you address your highest spiritual principle), and the answers will be more accurate. With the Protocol you will never be given more energy or information than you can handle.

The Dowsing Protocol I'm sharing with you below may seem like a lot of trouble to go through, but if you use it before each session, your accuracy will increase. If you do the first five steps of it several times a day, even when you're not dowsing for wells, you may find that you experience more balance in your life as you "clean up" the energies around you.   Feel free to adapt the Protocol to your own traditions.

This Protocol is a template that you can adapt to every situation by substituting a few words. When you go through it the first time at the start of a session, your intention is to prepare yourself for accurate dowsing (Steps 1 – 6).

In Step 6, focus on the specific issue you are addressing. For example, you could dowse for where the best place is for a driller to drill to obtain a sustainable source of good tasting water that meets all regulatory requirements and is sufficient to meet the needs of the client. Adapt Step 6 to the specific client you are dowsing for, including their specific requirements.

Allow the pendulum to swing freely as you go through the Protocol and ask it to indicate YES to show you that the steps are complete. If you get a NO response at any stage, you know you need to resolve, as best you can, that step before going on. For stubborn issues, use "Create an energy matrix" (Step 8) to reduce nonbeneficial barriers and increase beneficial influences over time.

Susan Collins' Dowsing Protocol

Email Susan for a free PDF of the Dowsing Protocol of the Summary: susan@dowser.ca

ALWAYS use a dowsing protocol before you dowse to help make your answers more accurate and keep you safe. Hold the pendulum in your hand while saying these words to confirm your answers. Use the first 5 steps of this Protocol to get YOURSELF ready to dowse. Then check with Step 6 to see if it is right for you to dowse on a particular topic at this time. If YES, then apply Steps 4, 5 and 8 to the issue to create lasting results. If NO, try again later, reword the question or try to work out why you are getting a NO response. When done, close the session (Step 9).

1. Balance your physical body (YES response)
Find a quiet time and focus your intention to balance and ground every aspect of your being.

2. Connect to your Dowsing Consciousness (YES responses)
Say: for the best and highest good of all creation (or however you address the Divine) I ask:
- To be connected to my human body for my good health
- to be connected with the intelligence and beneficial energies of nature
- to be connected and in resonance with Divine Good
- to be connected, guided and protected by my Spirit Team
- that my dowsing be 100% accurate

a. Check the tool's signals for YES and NO

b. Check the tool's signals for DISCONNECT and MAXIMIZE

c. Set your INTENTION: what do you want to accomplish in the session?

d. Ask for the assistance of beneficial energies in resonance with you and the Divine Source who have useful information to share at this time to assist you in gathering information so you can make decisions and take appropriate actions to achieve your Intention as appropriate within your Territory. (Your Territory is the physical and energetic area you have a right to work within.)

e. If you are working with a client ask that these beneficial energies also be in resonance with them.

f. Check for the presence of the beneficial energies you asked for in d.

g. Confirm that they are aligned with the Best and Highest Good.

3. Forgive yourself (YES responses)

We cannot be accurate if we have not forgiven ourselves and those around us.

Say: *"Creator, forgive me. I forgive myself. I forgive all those who have harmed me. I release them from my body mind and spirit. I ask all those who I have harmed to release me."* If you cannot forgive someone, give them back the accountability for their actions. You are a survivor not a victim.

4. Clear yourself of nonbeneficial energies

(DISCONNECTING responses)

With dowsing tool in hand say: *"For the best and highest good and as appropriate, I ask that all nonbeneficial energies (emotions, thought forms, attachments etc.) and processes (biomechanical, biochemical, bioelectric) associated with every aspect of my being (physical, mental, spiritual, emotional and energetic) be immediately removed in all dimensions, time frames, realities and frequencies including all nonbeneficial energies that are known, unknown, hidden, secret, stealth, disguised, fluctuating, fractal, reversed, residual and potential including all nonbeneficial cellular memories and energy triggers, including all nonbeneficial psychic cords and quantum connections."*

(If the tool continues to DISCONNECT after repeating the above three times go to Step 8.)

5. Maximize your energy field (MAXIMIZING responses)

"I ask that my energy field be maximized for the best and highest good of all creation and as appropriate and that all aspects of my physical, mental, spiritual, emotional being exist in perfect health in all dimensions, time frames, frequencies and realities and that I am guided in my thoughts, actions, and choices to be in resonance with the Divine Source." (If the tool continues to MAXIMIZE after repeating the above three times go to Step 8.)

6. Seek permission to dowse (YES responses)

If you get a NO to any of the following questions, do not proceed AT THIS TIME.

May I dowse for _____? (Do you have Permission?)
Can I dowse for _____? (Do you have the ability?)
Should I dowse for ____? (Is it for the best?)

7. Dowse

Use the processes described in Steps 4. and 5. above and apply them to the situation for which you are dowsing. Dowsing works best when you're in a state of ignorance and apathy: you don't know the answer, and you don't care what it is.

- Keep the question clear and literal.
- Assume nothing.
- Use a chart for accuracy.
- Respect others' privacy. Don't dowse unless requested to.
- Dowse in service for others, not for personal greed.
- Never diagnose or offer medical, legal or financial advice unless you are a licensed practitioner.

8. Create a matrix if needed (energy system) (YES response)

If the situation doesn't resolve itself within a few minutes, ask that an ever-changing energy matrix be established in the appropriate place, staffed by the appropriate beings, that will automatically adjust and transform all non-beneficial energies and processes to beneficial as appropriate within my Territory.

9. Disconnect (DISCONNECTING responses)

Fully, consciously, actively and as appropriate at this time, disconnect from all energies with which you have been working. (You may stay connected to the Divine Source and your Spirit Team and other energies that need time to "cook" for a while. Making sure you are disconnected from other energies will ensure you don't stay in resonance with them which could create fatigue, disorientation and even illness.)

10. Thank
Thank all energies and beings that have assisted you.

11. Communicate your results appropriately
If you are dowsing for someone else, be sure to check which results can be communicated to the subject for the best and highest good of all creation. Never discuss anything with anyone in a way that could identify the subject or create further problems. Be discreet.

Adapt this protocol to your needs by changing or adding any other words or prayers that feel right.
If you don't have time to go through it all, say something simple like: "Bless this situation." It's better to say a quick, simple prayer, even "BLESS YOU" at the moment it's needed than nothing at all because you don't have time to go through the whole protocol.

Please see Susan's other books for more information on other dowsing uses and techniques, available from www.dowser.ca and Amazon

Two thumbs up for a successful, new well.

Belief Systems

One of the main barriers to successful dowsing is the belief that we can't do it. The other belief that can hold us back is that we're not worthy of doing it. If you can suspend those disbeliefs, your dowsing will improve.

If you have trouble accepting that dowsing works, just act as if it works, and the results will follow. When I started out with well dowsing, I had to keep repeating over and over: *"I can do this"* Repeating this phrase became a mantra to shut up the part of my brain that didn't believe it is possible to find water.

Successful dowsing is based on need, not greed.

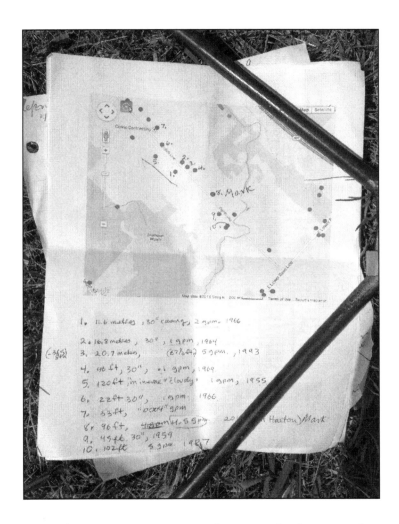

Before going on site, see if you can find a map of existing local wells indicating their depth and volume. Contact your local government for this information.

5   Dowse a Water Well

This is how I dowse a well. Other people may have different techniques. Talk to as many water dowsers as possible and develop your own method.

When you are contacted to drill a well, state immediately that you cannot guarantee your results, but you will help them if you can. Remind them that the decision of where to drill is up to them. You do not want to be liable for the drilling costs of a dry hole.

Set up the appointment to meet the owner, property manager or driller at the site. (Leave a note at home as to the location and contact information for the job.) It's best to pick a time when a work crew is not doing construction. If you arrive and a crew is working, ask if the generators and radios can be turned off for the time you're dowsing. Ask to have electric cattle and dog fences turned off so that those energies do not interfere with you. Don't attract too much attention to yourself, or you'll be dealing with spectators and having to block thought forms.

Chat with the homeowner to determine their needs, identify potential sources of contamination (such as a septic bed or storage shed), identify where the water currently comes into the house (so they can reuse their existing plumbing, if possible), and find out where they would ideally like to have the well.

Once you have that information, ask them to leave you alone while you dowse, and not to watch you or think about what you're doing. As you tune into the land, you may start also tuning into their thought forms, or the thought forms of other people who have lived there before. Ask your inner self to be free of any interference from persons living or dead. Ask for "The Truth".

Connecting with the Energy of the Place

We know that moving, underground water gives off an electromagnetic energy field that humans have the ability to detect. As dowsers, our bodies learn to read the energy frequencies, just as our eyes can tell the difference between colour frequencies (is it red or blue?) and our ears can tell the difference between audio frequencies (is it a guitar or a flute?).

Dowsing tools connect the body's knowing with our conscious mind so that we can determine location, depth and volume. I personally do this by connecting with the "Energy of the Place" – what I am really doing is bringing my mind/body into focused resonance with the energy frequencies of underground water. If the following doesn't suit you, then develop you own way of connecting with the land.

The planet has an electromagnetic characteristic that runs North/South. I begin my dowsing session by getting out my compass and finding the directions so that I understand where I am standing in relation to the earth's electromagnetic grid. It helps me focus on the electromagnetic signature of moving water under-ground to know what the background radiation is. I also assign metaphorical characteristics to each direction. East is the energy of beginning, South is the energy of doing, West is the energy of completion and North is the integrity of the project.

I face each direction, beginning with East, and using my L-rods, ask for the cooperation of those directions in completing the project. If I get a NO from any of the directions, I am alert to an issue in that part of the project. An example of this is a client that wanted to build a house and a trout pond in a flat, treeless, dry field. My dowsing response when facing West was a NO, and a few questions informed me that while the house might be completed, the pond could not be completed in accordance with the client's vision.

Triage Protocol

A triage method (from the French for "sifting") is used in hospitals as a way to quickly identify and resolve problems. I triage the various on site "Energy Levels" as a way of focusing my sensitivity and finding energy areas that may be resistant to the type of well or pond the landowner has in mind. One item at a time in the list below, I use dowsing to ask for the COOPERATION of the "Energy" of the:

- neighbourhood
- property
- house
- earth
- air
- fire
- water
- ancestors (these may be in the lineage of the current family or of previous people on the property).

- guardians (these may be ancient peoples who have a strong connection with the land that transcends their life span)
- animal kingdom
- plant kingdom
- drilling equipment
- "higher selves" of the homeowner, driller and crew
- all factors known and unknown.

If you get a YES (the energy will cooperate), proceed to the next level. If you get a NO, then make a note of that and go to the next level. **You must get all levels to a YES before you can be confident in your accuracy.**

How do you convert a "NO" to a "YES"?

As dowsers, we know we can communicate with all things. To get to "YES", we play "Twenty Questions" and figure out what the barrier is, then remove the barrier with dowsing, using Step 4 of the Dowsing Protocol.

Setting the Program to Find a Well Site

Once you have a YES from each level, say something like: *"For the best and highest good of all creation and as appropriate, I ask to locate a place to drill a well which will produce year-round, sustainable, pumpable, good-tasting water that meets government regulations for drinking water and will not adversely affect neighbours' uses."*

Say *"The area I am looking in is here"* – indicate with your eyes and arms the desired area (determined after consultation with the owner) and taking into account ministry regulations).

Say to your rods: *"Please point to the <u>best</u> place to drill to obtain the water needed by (for e.g., this family of four people; this horse farm with twenty horses and a stable and two houses; this machine shop which will employ about three people; or this water feature which will have trout living in it). Please point to the best place to drill, and cross when you get there."*

Let the rods point in a specific direction, then walk forward with the rods until they cross. Plant a flag or other marker that will remind you of the spot. Check: *"Does the flag indicate the best place to drill to get the water needed by (this family)?"* If not, move it, and confirm.

Once you have the site, ask *"May I now check the depth the driller will need to drill to get the water?"* *"May I now check the volume of year-round water available with a working pump system?"* Do both by "chunking" the numbers. For example:

"Is the depth between 0 and 100 feet?" NO

          100 and 200? YES

          100 – 150 NO

          150 – 175 YES

          150 – 160 NO

          160 – 170 YES

          160 – 165 NO

          165 – 170 YES

          165? NO

          166? NO

          167? YES

"Thank you. Is the volume between 0 and 5 gallons per minute?" NO

> 5 – 10 GPM NO
> 10 – 15 GPM YES
> 10? NO
> 11? YES

"Thank you. Will a well at this spot produce 11 GPM of good water if a driller goes down 167 feet and puts in a working pump?" YES

"Thank you. Please show me the direction of the stream." (The rods swing parallel to the stream). *"Please show the direction of flow."* The rods will swing to show that. *"Thank you"*.

Mark this information on the flag, and let the homeowner know what you have found. The homeowner will probably ask you to check a couple of other sites. Do that and flag them. Discuss with the homeowner what you have found and remind the homeowner that it is their choice where to drill and you do not guarantee the results.

I usually ask the homeowner to stand on top of each flag, facing their house, and ask them to feel the energy of the water under their feet. I then ask them which site feels the best to them. This helps give the responsibility of where to drill back to the homeowner. Some homeowners don't feel anything during this exercise, but some feel the water strongly, and I once had a client with no previous experience, who said she could actually "see" the lines of water.

Summary Water Dowsing Procedure:

1. "Empty" yourself.
2. Do the Dowsing Protocol. (I recommend doing this on the way to the job site so you are ready to go when you get there.)
3. Connect with the Spirit of the Place.
4. Set your program to find a spot to drill.
5. Dowse for location, depth, volume.
6. Plant a marker flag and double check that the flag is in the right spot.
7. Bring the owner back and discuss the spot. Mark up to about three more spots if asked.
8. Invite the owner to stand over each flag and "feel" the energy of the water. (Some will experience it.) Remind them it is their choice where to drill.

Map Dowsing

Water well dowsers will often map dowse a site before they walk the property. This technique is a quick and easy way to find out a lot of information without actually being in a place. To map dowse, you need an accurate representation of the property which could be a map, satellite image, photograph of the land, surveyors report, or simply a drawing with the address of the property written on the page. This image is used as a "witness" for the actual land.

You will also need a ruler, or other straight edge (I usually use my dowsing rod), and a pendulum.

Summary Remote Water Dowsing Procedure:

1. Do the Dowsing Protocol and triage the energy.
2. Determine what volume and depth well you are looking for. That is the Target.
3. Slide a ruler slowly across the picture vertically with your non-dominant hand, while holding the pendulum in your dominant hand. Ask that the pendulum give you a dowsing response when the straight edge crosses the Target.
4. When your pendulum responds, draw a vertical line at that point.
5. Now slowly slide the ruler horizon-tally across the picture and ask for a dowsing response. Draw a line.
6. Mark where the lines intersect. That's the site. Confirm the location when you get on site.

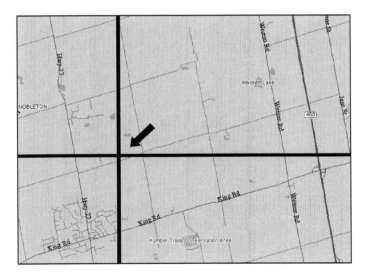

Stream Diversion

It is sometimes possible to actually divert an underground stream either to fill a dry well or pond or to move away from a wet basement in times of need. It is hard to explain the hard science behind this phenomenon, but the proof is in the pudding, as they say. Experienced dowsers have done it, either on site, or remotely.

Here's the procedure:

- Do the Dowsing and Triage Protocols.
- Check to see if there's a stream that's willing to move in the desired direction.
- Get a piece of "rebar", (a metal reinforcing bar used when pouring concrete) or something similar, and check with your dowsing where to put it.
- Get a hammer. Ask how many times to tap the rebar and in which direction, while asking the

water to move either into the well or out of the basement.

- Check with dowsing how long the results will take.

You can also do this on a map, using your rods in place of rebar and hammer. You can practice this technique with friends who also dowse. Have one person responsible for moving the stream, and the others can circle around the area, and use their rods to monitor as the stream moves. If you do this as part of a practice session, please be sure to put the stream back to where it was originally! If you do it in a real situation, be sure that you are not interfering with anyone else's use of the stream before you start.

6   Conclusion

It is not hard to learn the techniques for dowsing water wells, but it is difficult to build confidence without actually doing a real well. It will help your confidence if you know something about well regulations and construction techniques, so learn as much as you can before you begin, then simply trust yourself.

Practice your dowsing and do the Dowsing Protocol (page 34) and Triage protocol (page 43) before each session. Do as much research as you can before you go on site by looking at Well Records and checking aerial maps to see what the overall countryside looks like. When you get on site, make sure you know what your client is looking for and if there is new construction and where it will be.

Once you understand what's needed, ask your client to leave you on your own while you dowse. It's very distracting to have people following you around. Relax, triage the local energies (page 35), and let your rods guide you. Leave a semi-permanent marker in the ground and make a note of the depth and volume. If possible, also mark the site with GPS tags and provide those to the client and the driller.

With practice, you'll gain confidence and be able to locate good well sites and provide depth and potential volume of water. For me, there is nothing better than finding a good source of drinking water for a client. **Water is life!**

Appendix 1 Glossary

Annular Space: the area surrounding the well casing.

Artesian Well: a ground structure that puts enough pressure on an aquifer to move the water in the aquifer to the surface.

Aquifer: a water bearing layer of permeable rock, sand or gravel.

Aquitard: an impermeable layer of rock or overburden that stops vertical movement of water.

Bedrock: solid rock formation.

Bishops Rule: in dowsing, the depth of the water below the surface can be determined by the distance one walks away from the centre of the steam at the surface.

Bore hole: the hole that can become a well.

Buried Debris: potential barrier to drilling.

Capillary Zone: area above the water table where groundwater is drawn upward.

Casing: the well lining.

Dowsing Protocol: a way of accurately accessing your natural intuition.

Drawdown: the lowering of the water level in the well.

Drilled Well: constructed with a drill rig, often with a diameter of 7 – 15 cm (3 – 6 inches).

Driven Well: constructed by pounding the casing into the saturation zone.

Dug Well: excavated by hand or backhoe, often having a large diameter (2 – 3 metres / 3 – 6 feet).

Earth Energies: various energy systems in the earth, such as magnetic and electromagnetic.

Earth Faults: provide a pathway for underground water flow.

Electromagnetic: having both electric and magnetic characteristics.

Energy Field: region of electric, gravitational, magnetic etc. influence.

Formations: the type of geology underlying the surface.

Fracture: breaks in the earth due to faulting.

Frequency: how often an energy wave repeats in a measure of time. Measured in Hertz (Hz) or Cycles per Second (CPS).

Geomancy: the art of designing and placing structures in the landscape so that the Earth Energies enhance their intended use, and so that the structure itself is in harmony with the environment. Similar to Feng Shui.

Geopathic Stress: the impact of non-beneficial earth energies on living organisms.

Geothermal well: a well that uses the temperature of the earth to heat or cool water.

Glacial till: material (sand, rock and gravel) left by glaciers.

Groundwater: water below the surface of the earth

Groundwater discharge: when the water table rises to the land's surface, groundwater discharges into surface water.

Groundwater recharge: groundwater is recharged from precipitation - either rain or snow melt.

Hertz: Hz. Cycles Per Second (CPS). Used to describe frequency.

Higher Self: an unconscious, subconscious or super-conscious version of oneself.

Hydraulic Head: water level in a well.

Hydrofracking: clearing a well of debris by forcing pressured water into the well in order to increase flow.

Hydrogeology: a branch of geology concerned with the occurrence, use, and functions of surface water and groundwater.

Infiltration: movement of water into formations.

Jetting: propulsion of high-pressure water into a sandy aquifer to create a hole for a well point.

Karst Topography: sometimes found in limestone formations and marked by sink holes, little surface water and large springs.

Magnetic Anomaly: area with unusual magnetic characteristics.

Magnetic Images: false readings of earth energies due to magnetic deflection.

Map Dowsing: using a map to remotely detect the location of something.

Moraine: an accumulation of earth and stones carried and finally deposited by a glacier.

Negative: (-) refers to the electrical charge of an object. Does NOT mean non-beneficial energy.

Overburden: earth (often sands and gravels) overlying bedrock.

Perched water table: a pocket of water above the normal water table.

Polarity: the electrical charge or condition of a body. Positive (+) or negative (-).

Positive: (+) refers to the electrical charge of an object. Does NOT refer to beneficial energy.

Primary Water: water formed by chemical reactions in the earth.

Pump test: constant pumping over a period of time that identifies the flow of water available from the well.

Radiation: the energy transfer of electromagnetic waves from, for example, moving water.

Rays: lines of energy.

Remote Sensing: detection at a distance.

Resonance: responding to vibrations of a particular frequency, especially by itself vibrating.

Sandpoint Well: a well created in sand or gravel by driving a pointed well casing into the ground.

Saturation zone: the area of the ground saturated with groundwater.

Septic system: treatment system for household wastewater. Pollution from septic systems can include nitrates, bacteria and chemicals poured into the system.

Sink Hole: a hole created in the earth when rock, often limestone, is dissolved by water.

Porosity: the ability of rock to hold water. Gravel and sand have a high porosity. Limestone tends to hold water in cracks. Clay has low porosity.

Shock chlorination: a process of eliminating contamination in a well.

Spirit of Place / Genius Loci: In Roman mythology a *genius loci* was the protective spirit of a place.

Spring: source of water where the water table meets the ground surface. They will dry out if the water table drops. Hot springs are generally confined to areas of recent volcanism where ground water is heated through contact with magma.

Static Water Level: the level in the well when no water is being withdrawn.

Surface water: water at the surface of the earth such as wetlands, lakes, ponds, rivers and the ocean.

Thought Form: an energy pattern produced by thoughts or emotions that can remain as an imprint at the location.

Turbid: thick or opaque.

Vibration: rapid motion to and fro of an electro-magnetic wave.

Water Cycle: the continuous movement of water from the surface of the earth to the atmosphere through evaporation, and the return of water to the surface through precipitation.

Water Table: the top of the groundwater or zone of saturation. The water table rises and falls according to the time of year and annual precipitation.

Wavelength: distance from the crest of one wave to the next.

Water Dome/ Blind spring: primary water forced up under pressure towards the surface of the Earth. At some point it hits an impermeable layer of rock or clay and then moves laterally into veins (often five).

Water Table: the upper limit of the portion of the ground wholly saturated with water.

Water Veins: subterranean water flow.

Water Well: a hole that extends into the earth until it reaches an aquifer. Wells pump water from aquifers. Three basic well types are common: dug, driven and drilled wells.

Well Records: records prepared by the driller at the time of drilling that describe the well and ground.

Well screen: perforated cylinder attached to the bottom of the solid casing to keep solid particles out and let water in.

Witness: a sample of an object being sought by dowsing.

Zone of aeration: the area of the ground holding air.

Rev. Susan Collins
Management Consultant
Professional Dowser
Metaphysical Minister
Canadian Society of Dowsers
    Past President
    Dowser of the Year

Presenter at National Conferences
Alien Cosmic Expo: Brantford, Toronto, ON, Canada
American Society of Dowsers Conference: Lyndonville,
    Vermont; Saratoga Springs, NY, USA
ASD West Coast Convention: Santa Cruz, California, USA
ASD Southwest Conference: Flagstaff, Arizona, USA
Binnaji General Trading Co: Kuwait City, Kuwait
British Society of Dowsers Conference: Cirencester and
    Leicester, UK
Canadian Society of Dowsers Conference: Toronto, London,
    Markham, Peterborough, ON, Canada
CanAm 1, ASD/CSQ: Harrison Hot Springs, BC Canada
Questers Conferences:100 Mile House, Salmon Arm,
    Harrison Hot Springs, BC; and Olds, Alta Canada
Foundation of Mind Being Research: Palo Alto, CA, USA
International Dowsers: Sterling, Scotland
Italian Dowsing Society: Bologna, Italy
Japanese Society of Dowsers: Kakegawa, Tokyo, Japan
Ozark Research Institute Conference: Fayetteville, AR, USA

Many regional meetings and events across America.

Author: Books (Kindle and Print)
Bridge Matter and Spirit with Dowsing
Get Healthy with Dowsing
Dowsing for Feng Shui and Space Clearing
Meet Alien Energy with Dowsing

Meet Orbs with Dowsing
Use a Protocol to Get Results
    Classic, Bible, Muslim and Japanese editions
Dowsing Triage – Finding and Fixing Energy Problems
Water Wells – What a Dowser Needs to Know
Get Happy with Dowsing – Change Unhealthy Patterns
Life Cards Oracle Set

Workshop Rentals
Susan's curriculum is available on Vimeo.
https://vimeo.com/susancollinsdowser/vod_pages

Publications
Anemone Magazine, Japan
What's New in Dowsing, CSD journal, Canada
The Quester, CSQ/CSD quarterly journal, Canada
The American Dowser, ASD Quarterly Digest, USA
Dowsing Today, British Society of Dowsers, U.K.
Journal, The Dowsing Society of NSW, Australia,
Revista Cientifica Radiestesia, Dowsing Society of
    Chile
Human Spirit Magazine, Ontario, Canada
Vitality Magazine, KI Awareness – Ontario, Canada

Film and Television
Resonance Film
https://vimeo.com/ondemand/theresonance

City TV; Rogers TV; VRLand News.

YouTube Channel
www.youtube.com/c/susancollinsdowser

Susan Collins Print and Kindle Books

Check Susan's Kindle and Amazon Print on Demand books in the "**Complete Guide to Dowsing**" series.

Print books by Susan Collins:
   Bridge Matter and Spirit with Dowsing
   Get Healthy with Dowsing
   Get Happy with Dowsing
   Dowse for Feng Shui and Space Clearing
   Meet Alien Energy with Dowsing
   Meet Orbs with Dowsing
   Water Wells: What a Dowser Needs to Know
   Life Cards Oracle System

To order books and tools, or to arrange a workshop or personal session, contact Susan at susan@dowser.ca www.dowser.ca

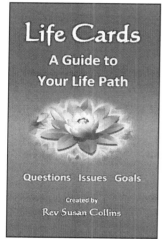

Life Cards Oracle Set

Rev. Susan Collins is an internationally acclaimed dowsing teacher, Keynote Speaker and workshop leader and is an ordained Metaphysical Minister recognized by Service Ontario, Canada. She uses traditional dowsing tools as well as the power of heart and thought to detect and transform nonbeneficial Earth, Environmental, Psychic and Other energy patterns.

She has presented in person at many international and regional conferences across North America, the UK, in Italy, Japan and in the Middle East and was featured in "The Resonance", a feature-length documentary on extraterrestrial presence on our planet.

Susan has a dynamic, global consulting practice and is the author of a successful series of books.

She is a Past President of the Canadian Society of Dowsers (2003- 2006) and was named Dowser of the Year in 2006.

Authors note: please see my other books and rental workshops for more information on how to use dowsing for self-healing and balancing energy or to book a personal session.

susan@dowser.ca

www.dowser.ca
www.InternationalDowsers.org

Printed in Great Britain
by Amazon

32442896R00036